A PRIMER OF KINETICS

A PRIMER OF KINETICS

James L. Rosenberg

NEW POETRY SERIES

ALAN SWALLOW, *Denver*

Library of Congress Catalog Card No.: 61-9436

CONTENTS

I: *Animae*

II: *Personae*

ACKNOWLEDGMENTS

Acknowledgment is made to the following magazines in which these poems previously appeared: *A Houyhnhm's Scrapbook, The Antioch Review, The Atlantic, Best Articles and Stories, The Borestone Mountain Poetry Awards, Epoch, New Orleans Poetry Journal, New York Times Book Review, Prairie Schooner, PS, University of Kansas City Review, Voices, Western Humanities Review.*

I: *Animae*

A CHILD'S BESTIARY

1.

A tree is a tall fellow,
A motionless dancer;
He moults in the winter-time,
And in the spring
Bird's-nests, fruit and nuts
Drip from his spiny fingers.

2.

A stone is a still person,
At his core:
Such brightness, tightness, coldness
Mind cannot utter.

3.

Water goes clear ways on invisible feet
In soluble freedom;
When cold comes,
It gathers to invincible stillness.
Warm nights, under bright stars,
It smiles in its dreams.

4.

Birds are a sense of float and swoop
Made manifest in feathers,
And their voices:
Spirits of little bells that float and swoop.

5.

The snake is a soundless sinuosity that parts the grass;
Dire and sharp and bitter his little tooth!

6.

The butterfly is a lace of dancing jewelry;
He lives a sweet but brief moment in blossoms.

7.

The sky is a giant emotion;
Sometimes blue with happiness;
Sometimes terrible in tears;
Sometimes electric with rage;
Sometimes snowing benedictions of bright crystal.
He has no beginning or end.

8.

The fish slips cold ways in wet latitudes;
Lithe as a current of flowing
And turned in the lathe of the water,
His life is a long curve.
Cool in his running silence, he speaks a language of
bubbles.

9.

The snail lives in a shut house.
Testing the texture of his town with shrinking prongs,
He draws back in upon himself when hurt,
Seals up his valves,
And thinks zero.

10.

What a man is, I do not know.

CARDINAL UNDER GLASS

Grandmother's sitting-room was neat
And always dusted behind drawn blinds;
Except for Sundays, no one stepped
Within that pious ambience,
That core of silence in the heart
Of larger silence, in the heart
Of circling, silent wilderness,
The gray, neat, clean and pin-sharp house
Encircled by a palisade
Of geometric hollyhocks,
And in that noiseless sitting-room,
As antiseptic as a church,
There burned beneath a bell of glass
A stuffed and bead-eyed cardinal:
His plumage glowed like smouldering blood
Within that room's Victorian hush,
A core of burning in the cool;
Beside him, on the tall buffet,
A foursquare, black-bound Bible stood,
Hasped with a mighty lock of gold;
Above these on the papered wall
Grandmother and grandfather hung,
Framed in solid, sensible wood,
And staring severely down at me
Where I crouched through the lonely hours
In childhood's demon-thronging woods,
In silent adoration of
The stiff, stuffed cardinal under glass,
Burning, burning in the gloom
Where the correct clock coolly ticked,
Primly ringed by hollyhocks,
Ringed in turn by wilderness

That knew no proper bound or law,
And all ringed in and circumscribed
By vast revolving wheels of fire
Beyond the sitting-room's decorum.

BEE STINGS CURE ARTHRITIS

My swollen fingers, gnarled and knotted,
Cramped awkward with intolerable ache,
I bought a farm of snarling bees;
They stung me for salvation's sake.

Strange ease, to purchase through small hurts
Release from larger, crippling pain!
But I have learned a strange adjustment
And count each fiery sting a gain.

And sometimes in the savage orchard
I go to take my hurtful ease
And find a comb of gleaming honey
Left by my fierce redemptive bees.

A BEDTIME STORY: FOR MY DAUGHTER, AGED SIX

The wolf who lives in printed pages,
Half Iago, half slapstick lout,
Bellowing his improbable rages,
Hoarsely bleating, obscenely wiggling
Under his feigned and fleecy pelt,
Evokes an easy evening's giggling
And gooseflesh squeals at his bawdy evil;
And I, as I read his escapades
Out of the flimsy, gaudy book,
Fake fear too and shake like the devil,
Secretly knowing the turn of a page
Will vanish him in a puff of smoke.

How well it would be if such were the truth
Of the real wolf, circling our house in the night,
But I cannot warn you, I cannot preach
The physical fact of his claws and teeth
Or his fetid breath, smelling of hate.
There are some truths no talk can teach.
But the pages will turn, and time will pass
And slyly seduce you out into the dark
To a rendezvous in a horrible park,
Where wolves and serpents couch in the grass.
Gladly I'd barter my chances in Heaven
To shield you from all the world's wolves, oh my child;
But such dispensation has never been given.
Take care. I love you. Go walk with the wild.

THE DESTRUCTIVE ELEMENT (ON TEACHING MY DAUGHTER HOW TO FLOAT)

"Fear death by drowning"—older than the race,
Written in our nerves; we shrink
From water's cold ambiguous embrace,
And come up choking. Water is to drink,
Not breathe,
And yet:
That swimmer soonest sinks who fears
The water's love, my love.
Here on my arm
You lie, with water whispering round your ears,
Your body tensed against the liquid harm
Of nothingness that seeks to suck you down
And strangle you.
But water does not kill.
It is the fear, the fear that dooms us down.
Oh, give yourself to the water, fear no harm,
And after my impermanent, sustaining arm
No longer's here to steady and enfold you
Lie easy and accepting in the water's love
And it will hold you.

SERENADE FOR DOVES

This night
Beside the pallid light
Of my sad desk, I sit and write
To you,
Who two by two
In branchy darkness snuggle and rou-coo
In keeps
Of cedar, and in holts
Of oak; who snuggle featherly and warm, in love.
Out there,
Haloo!
I sing you.
And I think
How your loves mock
My warped pen and my sludgey ink,
Mock and make false
The spent and feeble force
Of this, my one-winged fluttering verse.
Oh, living loves,
How your bright commingling proves
The poverty of my weak wobbling words.
But art is short
And love is long.
Light me. Be my song.

IN MOTH-LIGHT TIME

In moth-light time, in humming dusk,
The air, a lavendering smoke,
Gathers around the trees like a lake,
Alive with ambiguities
That flutter and dab their colorless wings
In minuscule, disturbing clashes
In the cold grate. I rock and creak
In my pool of artificial dark
(My vine-enclambered porch) and peep
Myopically through the watery light
Where all shades flow and run together,
And trolls that yesterhour were trees
Stretch and clutch with twiggy claws;
A million enigmatic gleams
(Diamonds? Fox fire? Little eyes?)
Flash from the crouching, neutral lawn;
And overhead through floating darks
There wander vaguely like drifting souls
Strangely muted eldritch shrieks.

Whoever you are, who are floating and calling
Out there in the misty, murmurous dark,
Stand clear and let me feel your name!

No answer out of the melting night.
I go inside, snap on the light.
The naked bulb floods all the room
With unambiguous brilliancy,
In cynic impassiveness revealing
An unpainted chair, a cracked ceiling,
Plain and ugly and solid as sin.

Yet comforting. I sip my beer,
Sigh and relax. There are answers here.

THE TOP

The top in its latent spinning hums and stills
To vibrant sentience, to molded motion.
Its life is movement, and it fills
With perpendicularity
A possible space; it fills with fusion
Of what has been and is to be
The pregnant emptiness of inaction.

More-than-a-top, its life is motion
(Its death the cessation of sentient spinning);
It only exists in motionless action,
Assuming the change of form in turning,
Which gives all whirling worlds their sanction.
It dies in wobbling to a stop—
Drops inert—a chunk, a block.

CAT AND MOUSE

On a bright May day
I watched a lovely cat,
Lithe as a velvet muscle,
A-killing a field mouse
On the lawn before our house.
Under the bonny, smiling sun,
With the air as sweet as breath,
I watched upon my emerald lawn
The gracious dance of death.
Oh, its feral choreography
Pleased my critic's eye:
The cat in flawless flowing
Of living fur and claw,
Wise, alive, disinterested,
Death without a flaw;
The mouse, stunned by immensity,
But firm committed, too,
To playing out the drama
(Since what else is there to do?).
The cat's eyes burned like opals,
Her every move was music;
The mouse's eyes were round and great,
Not with fear, but wonder,
And, staggering, he still maintained
His ritual refusal
Of the deep and lovely potency
That raked him with its talons;
He feebly hopped, from his tiny nose
There hung like a little ruby
An infinitesimal drop of blood,
He kept his precarious balance.
(Though wisdom might dictate
A prudency of submission
To incomprehensible fate.)

How long the game continued,
I cannot rightly say—
A bit too long for me, I fear,
Or for the mouse's pleasure,
Thought just about appropriately
For that cool and golden leisure
Which activates the cat
With its rhythms of renewal,
Those rhythms which are neither bad
Nor good nor kind nor cruel,
But simply *are,* as God is.
(If, that is, He *is.*)
And now as I sit and write
My cat lies at my feet
Humming her warm domestic love
And honing her velvet jaw
Against my civilized shoe,
While over the lawn the stars
Look down from the sweet night skies
Like a thousand thousand thousand
Opalescent eyes.

II: *Personae*

SON SONG

My father lived his faith in Ford
And read the Saturday Evening Post
Religiously, and he adored
Those packaged cereals by Post
Which made our nation what it is: revered
Around the rolling world
For Yankee know-how; the flag unfurled
In the Saturday newsreels made him puff
With patriotic paternal pride,
Though even then this was enough
To sometimes make me cringe inside
(And less inside as years went by).

What you eat is what you are,
They say. For years I feasted on
Modern America, Middle Class,
Without a let-up, without a stop:
Hot dogs, apple pie, and Mom
And gluts of fizzy, sweetish pop—
These built my bones and formed my flesh.
Anatomy is Destiny,
As sure as God made good green cash.
Trapped in flesh I never willed,
How hopelessly I try to unbuild
The force that fired my senseless clay;
But General Electric will sing in my brain
As long as my current of thought holds out,
And my only redemption lies in forgiving
The fate that shaped me as I am;
I must somehow forgive and go on living.

You, child of my blood and brain and bone,
You must do the same; you are not alone.
Every father murders his son
In an endless cycle of suicide
And the dead walk lone and far and wide
Looking out of their sightless eyes
Crying aloud: "Forgive! Forgive!"

Boy of my loins, my helpless victim,
Forgive those Fords by which I live.

AUTOBIOGRAPHY

My parents bred me where the winds were warm
And fed me fat on honey, figs, and dates;
They meant no serious conclusion, meant no harm,
But thought a baby might be fun to raise.
I grew, as simple as a chimpanzee,
To splendid ignorance in their unknowing hands
Beside the shining shores of that far sea
Whose rhythms still surge deep within my bones.
Half mystified, half angry at their luck
For having spawned such a recalcitrant odd fish,
They clothed and taught me, set me to my books.
I learned the Will is father to the Wish.
Baffled to sullenness, I read and wrote
Such stuff as trash is made on (and still do).
The passing years swelled slowly and grew ripe.
I did not really ripen, but I grew.
So silence built like coral in our hearts,
Triple strangers under a single roof.
We lived our days and mimed our separate parts;
I learned indifference cuts deeper than reproof.
A day came when I left that luscious place
And wandered over the miles and through the years
And now I live in a land of heat and ice
Far from childhood's palmtree-shaded shores.
Yet sometimes still I think of that bright garden
Where silence bloomed as rich as the winter flowers;
I think of all the syllables unspoken
That might have gone to fill those empty hours.
Here in this room my children run and play
And I cannot find the words or signs to tell them
How thoughtlessly I got them one blue day,
For there are things no one can say to children.
Oh, all the words that never can be said,

The love that somehow cannot be expressed
Burns brightly down the beaches of my blood;
My thoughts turn lamely toward the lonely West.
The land I write from now lies locked in snow,
A glassy silence dreams upon the streets;
God grant that some day soon warm sap may flow
And free the mute buds burning in the trees.

NOTES TOWARD AN EXAMINATION OF THE FUNNIES AS AN ASPECT OF AMERICAN CULTURE

How nice life in the funnies is
Where simple, unambiguous
Issues burst in asterisks
Of "Pop!" and "Bam!" and "Zowie!"
And though the colors sometimes bleed
Upon cheap pulp and blur the lines
The labels always make quite clear
To those who yearn to be made wise
The good and (bop!) the bad, bad guys.

In that bright world, grotesque enough,
Garish enough to clarify
What often in real life is dim
There are no problems, none at least
That can't be settled by fists and feet
And "Whap!" and "Bang!"; this certainty
Is bought, of course, at the expense
Of caricatured exaggeration
And slight adjustments of the truth,
But nothing's perfect, our sages say,
And it's pleasant at the end of a day

To lose oneself in the funny papers
And bask in sketched simplicity,
Relaxing the terrible vigilance
Required in the world of eight-to-five,
Required to merely stay alive
In a grey and complex world of wit
Where no one wears an easy name
And good guys, bad guys, merchants, thieves,
All go unlabelled, all look the same.

POST OF OBSERVATION

Man at a high window, looking down
On a world made little by a trick of space,
High as Gulliver over the Lilliput-town,
See: there under the lamplight—a white face,
And an arm that is moving mechanically up and down.

A man with a tire-iron is beating a dog to death.
The scene is etched in sharp lines of no sound.
The glass takes care of that. It shows your breath
When you sigh against it, but no noise comes through.
There, clear as crystal: you watch it against your will.
And there's not a thing in the world for you to do.

For all the good of your strength, you might as well
Be floating bodiless out in ethereal space.
There's nothing anyone ever can do or say
Against those things that happen beyond this place,
Though some will weep and rage and some will pray.

Nothing to do but helplessly hate what you see
And draw the blind against the evil source
And, learning the meaning of wisdom, turn away
To keep your new-won weakness warm indoors.

THE OLD KING'S SONG

Within the land that I fare from
I was a mighty king;
I rode a horse gold as the fire
And wore a blood-red ring
And hunted after the nimble deer
Through the green woods of that land
With twice two hundred baying hounds
And a long bow in my hand.

Don't weep that I am old;
I've known hot joy and rage;
Within the land that I fare from
I lived a pretty age,
In gleaming mail and armor
And blood-red velvet coats,
And falcons whizzed from off my wrists
Like Vulcan's thunderbolts.

Oh, I jingle-jingled as I rode
My fresh green countryside,
And all the folk bowed down to me,
And a queen rode at my side.
And love lived in that youthful land
And music and romance;
The days were sweet with sportiveness,
The nights with song and dance.

But greenest trees grow grim and brown.
The tenderest leaves grow brittle;
And man at last must turn him home
And rest from love and battle.
Now in my tiny garden
I live past longing's reach,
Surrounded by those little deaths
That make the garden rich.

THE GREENHOUSE AND THE FOREST

The greenhouse fronts the forest
With a regard of ice;
Its rectilinear glass
Rebukes the tangled trees
With a cool disinterest.
Within, its tender green
In regulated rows
Straight as a ruler's edge
Correctly buds and grows,
Exemplum for the forest
That like some shaggy beast
Creeps round and round the greenhouse
And shakes its rough green fur.
When the wind is in the east,
It moans and groans and stretches;
The greenhouse gazes coldly
As the forest snuffles and scratches,
But its diamond disdain
Will lessen in the moonlight
That whitely chills each pane
And makes it ache with fear
As, active in the moonlight,
The forest rustles near;
Bright though the greenhouse sparkles
In the strict eye of the sun,
The forest is implacable;
Its green will will be done.
And after the sparkling greenhouse
Has fallen into ruins

And our artifacted world
Has gone up in white flames,
The forest is forever
And shaggily will crawl
Over heaps of broken glass
And warped cucumber frames.

A PHYSICAL SONNET: ON A STUCK WINDOW

Who does not know the helpless thrust
Of flesh and breath, of mind and fist
Against the brute indifferent weight
Of things that are sullen and dull and mute—

Doors that are jammed and windows stuck,
Gears that strip, glasses that crack?
(Lord, what passive malevolence lies
In a world of objects wedged edgewise!)

The man who has never hurled his heart
Against the warp, across the grain,
Knows little of depth and less of dark;

But he who has, cursing, fought and faced
The dull-edged malice of jammed, sprung things
Can testify to the weight of hate.

THE JUGGLER

He who in sheer perverse diversity
Of nimble-fingered flip dexterity
Can keep
Eight balls spinning on the dazzling air
In pleasing and miraculous parabolas
Wins wonder from the fumble-fingered many
Who
Can't juggle one.
But those who clap, who flap flat palms,
Stamp, whistle, cry: "Hurrah!"
Do they not sometimes find
A something silly in such slick shenanigans
What skills it, after all,
To keep a plate, an Indian club, a ball
Spinning in perpetuity
Unless
Such leaping objects group and shape
To meaning
In some possible viewer's sight,
Unless
Some touch of art transforms
That flying bric-a-brac in glittering motion
Into a stilled, significant constellation.

MEDITATION IN A THREE O'CLOCK LIT SECTION

My youth was bright but misdirected
By dried old maids and lying parents
Who bored me through twelve years of schooling
And tried to tell me that appearance
And reality were one.
Oh, I was dumb—

But not that dumb.
I said, "Yes, ma'am," and smiled in secret
When dry Miss Abercrombie squeaked
Her squawky chalk aslant the blackboard
And within the drowsing classroom
Fat blue flies went buzz and buzz,
Sickening against the windows,
While the chalk screaked "Hiawatha."

Now I stand before the blackboard,
Proper in a business suit,
And my dry chalk whispers "Prufrock,"
While I think of that bored kid
Who somewhere somehow still is sitting
In a warm and buzzing classroom
Back down hallways of becoming
In old environs of the sun.

A GRIM FAIRY TALE

Simon, simple as the day was sweet,
His mother's youngest, and the family fool,
Awoke one midnight from a dream of cheese,
Strapped on his rucksack, sword and horn, and went
Swashbuckling into the blue and moony woods
Where froggy creakings lurked and small eyes blinked
And twiggy spooks clutched at his feathered cap.
But oh, so pure was his simplicity,
Simon never felt a tickle of fear,
Stopping only from time to time to knock
At some rude forester's embowered hut,
There to inquire for any neighboring ogres,
Marauders, trolls, bad biters, chuffing dragons.
But ah, alas! no work seemed in those woods
For hopeful unemployed young dragon-killers.

One day his way wove deeper through the gloom
(What bright birds chimed, bells rang, bones sang
Within that creepy and enchanted grove)—
He came unto a steaming, hissing cave;
Within: most monstrous dragon of the earth,
Rumbling, belching most sulphurously!
Did Simon hesitate, No thinker, he.
His sword flashed out: Snick-snack and snicker-snee,
And lopped the bellowing, cow-like head
Clean as a melon off its thick green neck.

When Simon, fortune's darling, some years later
Emerged at last from out those ringing woods
There dangled at his belt like bouncing fobs
Ten ogres' heads, eight scaly dragons' tails.
What homecoming was there, my friends,

What bands and banners, grand shenanigans,
Keys to the city, pompous speeches, songs!
No need here to recount at fatal length
How Simon, hero, triumphed after that,
Wooed, won, and wed the princess of the land,
Grew fat and prosperous with office, waxed
As do rich pumpkins on October nights.
Grown thus to greatness, his remaining days,
Plenteous with children, burdened with increase,
Slowly, gradually diminished like the light,
And leaves that once grew tender, green, and sweet
Hardened, reddened, in a sharpening weather.

Need more be said? Simon, half-forgotten,
Lived on a little while on memories
And age's thin-soup diet of old clippings
Pressed dry as leaves within a dusty album,
Grew weary then; unnoticed, one day died
Into that windy land where old bores go.

ANGELS UNAWARE

In the drear pictures of the Dutch masters,
How the dull commonplace takes place.
See, here's the colorless and lumpy side of life:
The housewife's truck and gear, and solid husbandry.
In the monotony of rural rounds, cups break,
Hems rip, bad cats spill milk, and dirt appears;
The good wife scrubs and mends, men plow
And smoke, some whittle thriftily, some bleakly smile;
One drily pares an apple in the sunset's glow.

All this lacklustre factuality
Looms dim and seems depressive to the eye,
Unless the mind can steadily conceive
(Above the picture and outside the factual frame)
Squadrons of seraphim in shafts of sun
Trumpeting their gold and silver alleluias,
Testifying how the Lord loves lowliness,
How strangely and unseen he showers down
(In benedictions our dull senses fail to apprehend)
Upon the greyly unrequited of this earth
His love. In ways unknown to us, these pictures seem
to say,
They also dully, doggedly do serve
Who scrub and plow and keep the narrow hearth.

MOVING LIGHTS

Boywise once on a rickety pier,
I saw, as night drained down the bay,
Two moving lights that floated on
The burning edge of the dying day—
Two moving lights: a phantom ship
Sliding silent as a cat;
The sucking water round the piles
Was whispering with a wet flat slap.
Those moving lights like floating thoughts
Rode lightly through the whispering dark,
As enigmatic as God's smile;
The night bloomed black, with looming stars—
These stars that here above by head
Trace their ancient patterns tonight:
Still moving lights, traversing still
The wavering dark of my dim sight.

GRANDFATHER DIED FOR EIGHTY YEARS

Grandfather died for eighty years
Powerfully toward timelessness;
In the far fields he waged his wars
Against an insect wickedness.
From our grange windows we looked forth
On his heroic husbandry;
The orchard trees bore at his touch,
The corn bloomed through his sorcery.
His holy thoughts were fire and flint;
He had no fear of faint or fail;
But all his love and strength were bent
To shape the life of significant soil.
Sabbaths he forged to church to bear
The awful weight of God's stern love;
He sang of endless evermore,
Yet all his work was grow and live.
The seasons were his almanac,
The stars ticked off his cyclic days,
And when it came the time to die
It filled him with a fierce surprise.
His mighty dying shook the house;
Eighty years of strength and growth
Are not too easily torn loose
From fruitful soil by the hand of death.
As the sea clears deeply, drawing out,
The house was drained of vital force;
So strong a letting go is not
Accomplished free of pain and stress.
At last the windows closed their eyes,
The candles had no more to burn;
Then silence moved upon the trees
And eternity on the corn.

THE HERMIT

Armed with God and an axe,
He entered the virgin forest
And made himself a clearing
Among the breathing trees
And a little hut in the clearing
Which silently sent up
A still plume of blue smoke.

But blue nights in the forest
The druid trees would steal
Insensibly upon him,
Obliterating the clearing,
And, hack as he might with his axe,
The passive might of the forest
Kept reasserting growth.

Then the axe's edge grew rusty,
And tentacles of despair,
Green and lithe as serpents,
Choked the hut in the clearing.
Now, if you walk in the forest,
Sometimes you can hear
The feeble, muffled strokes of an axe
Lost in that pulsing green,
And you see a strangely-colored bird
Flying among the branches,
Lost in the forest immensity
And screaming in despair.

MY SON

I see my blazing boy on fire
In the setting sun;
I see him flaming in an old desire
Of the son.

My sun is setting, and the night
Draws deeply on;
But he is on the rise, and my dying light
Becomes his dawn.

World without end of suns and exploding stars,
All fired by the same old rage;
The love that burns on Venus and on Mars
Warms my old age.

My son, my source of light, burns bright;
The light that once was mine is his;
I bid you, boy on fire, good-day; tonight
We'll meet where the Sun of Light forever is.

A SHORT HISTORY OF LIGHTHOUSES

Lighthouses, in the world's light-hearted youth,
When man could be an island if he chose,
Stood flashing their pinpoint gleams of rectitude
From Thebes to Tyre to Troy, where the castles and
    rocks
Lay picturesquely stacked like children's blocks.

Lighthouse keepers, too, in that silvery age,
Lived in lonely cragginess in their tall towers,
Calmly observing the ocean's measured rage,
Meditating the midnight moon's white powers,
Or gravely plotting the oceanic stars.

Now radar flips its serio-comic blips
From point to point in endless staccato clatter
Of spots and flashes of light and squeaky beeps
Of rackety sound. Farewell to silent weather.
The world is drowned in a rain of electric chatter.

The signals fly like sparks around the sky;
Yesterday's lighthouse-tower is Babel's today.
The world where a man could tend his private light
Is with Atlantis under the rolling wave.
How Donne must turn and smile in his island grave!

WORLD ENOUGH AND TIME

What happens to the clocks when time runs out
And the sifting sands of sense all leak away?
The watch maintains its mathematical pulse
Here in this timeless room this joyless day;
But soon there'll be no one living to translate
Its pure geometry of thought to syllables
And, when time dies, all clocks must lose their minds,
Exploding in a crystal shower of wheels.
Outdoors, grey light drains down from the livid air;
An emptiness appals the darkening sky.
One lone crow gives a single rusty squawk;
The ironic church-bell clanks in blank reply.

THE MAN OF GOD

The preacher at the party
Was newly come to town
To undertake his cure of souls
And when my turn came round
To say hello, he greeted me
With a handshake like a vise
And a smile revealing straight, strong teeth
As white as polar ice.
He tried a literary gambit, and
Asked me how I rated
Some of the modern writers.
I paused and contemplated
How best to slant my answer;
At last I mentioned Frost.
His amiably empty smile
Revealed that he was lost.
I said Shaw, too, afforded me
Considerable delight.
He asked me if I'd ever read
The works of Harold Bell Wright.
A silence fell. *The Robe,* he said,
Was the book of the century.
I smiled a trifle wanly,
But forebore to disagree.
Then, warming to his subject,
He apostrophized the Bard,
Recalling how he'd memorized
In high school, word for word,
Some splendid speech from one of the plays,
Beautiful but deep;
He said he *liked* deep stories.
I felt my foot asleep,
But nodded, murmuring assent,
With the tact of agony;
Just then a flowered matron
Chanced by to rescue me
By asking him to vouchsafe
His sanctified advice
On some sticky point of doctrine
Particularly nice,
I. e., could a man of middle years

Who'd never been to church
Have hope of his soul's salvation
Or would he be left in the lurch
On the day of awful judgment?
As they slowly moved away,
And I contrived escape at last,
I heard him richly say
That he was confident that God
Omnisciently foresaw
All possible exceptions
To the letter of the law.

How sure are his defenses
Who cannot comprehend
The nature of the enemy,
The measure of a friend.
"Sancta simplicitas," I sighed,
And I blessed him unawares,
Then ditched my cup, made my farewells.
And scrammed down the back stairs.

INCIDENT OF WAR

A head popped over the hedge;
I squeezed the trigger.
We went and found him lying in a sprawl,
A blond and pimply kid, I swear no bigger
Than a high-school boy.
The papers in his cracked and wrinkled wallet
Identified him as a corporal named "Krall,"
First name "Willy," birthplace "Dusseldorf."
My shot had caught him cleanly in the eye
And so his face was now not as God wrought it
(The other eye was open on forever,
A filmy, glassy blue, like a cloudy aggie).
Further search of his wallet then disclosed
A snapshot of a plump and homely lady—
"Mama," surely; on the homey wall
Behind her head there hung a painted plate—
"Der Herr ist meine Hirt"—and next
On the wall, a crucifix,
Whose Willy wanders wider ways today
Than that warm, spicy German kitchen dreams.
I knelt there in the brown, indifferent mud
And looked and looked my fill
On the subtle little hole that had let life out,
And then I gave the body back its belongings—
A picture, bits of paper, a faded name—
And shouldered the brute, indifferent weight of my rifle
And walked on through the dying of that day
Toward the years and years of nothing left to say.

GRAND CENTRAL STATION: CHRISTMAS WEEK

Perpetual arrivals and departures
Every hour at all gates, and hark!
From somewhere in the great dome overhead
A baritone is trilling like a lark:
"God rest ye merry, travellers!" A voice
Of brass proclaims some unpronounceable
Incomprehensible intelligence. "Rejoice!"
The baritone proclaims. The redcaps wheel
In planetary orbits, and we hear
Trains are leaving hourly for everywhere,
Sliding forth on gleaming ways of steel.
Here possibilities are endless and eternal.
What destination would you like? It's yours
For the ticket's price. Above the milling murmur
The baritone is praising what endures:
God's love, that is. But hurry, all aboard
For Hackensack, Infinity, and Cleveland!
Soon enough we'll rest and praise the Lord.
Now we praise the joy of pure existence;
Restless as a swarm of molecules,
Fussed by luggage, we are drunk with distance,
And must be moving on in endless cycles
Of departures and arrivals and survivals.

THE DARK WOODS

Strange woods along the river's edge
That used to sing me strangely to sleep
After-dinner under the trees
When summer nights grew soft and deep,
I hear you yet; I see us yet—
In memory's eye, in memory's eye—
A family group, the rich cigar
Grandfather smoked, the far-off cry
Of whippoorwills against the night,
The stealthy night across the lawn
Moving lithely under the trees,
Subtly, invisibly coming on
On leopard feet that leave no sign,
The maiden-aunts, the cousins, all
That Matthew Brady gallery of
Formal poses on the lawn—
Where are they now? All gone where love
Has gone to keep the darkening years
In the inevitable night.
A night-bird cries a hollow "Who?"
The stars are distant, cold and bright.
I, a little outside the group,
Seated, hugging my bony knees,
Am staring dreamily over the river
At those dark and distant trees—
And listening, across the river,
Over the years—and still I'm dumb
To answer. Woods, be patient a little.
Be still. I hear. One day I'll come.